our Environment

Oil Spills

Peggy J. Parks

KIDHAVEN PRESS

A part of Gale, Cengage Learning

GALE
CENGAGE Learning

Detroit • New York • San Francisco • New Haven, Conn • Waterville, Maine • London

© 2005 Gale, a part of Cengage Learning

For more information, contact
KidHaven Press
27500 Drake Rd.
Farmington Hills, MI 48331-3535
Or you can visit our Internet site at gale.cengage.com

LIBRARY OF CONGRESS CATALOGING-IN-PUBLICATION DATA
Parks, Peggy J., 1951–
Oil spills / by Peggy J. Parks.
p. cm. — (Our environment)
Includes bibliographical references and index.
ISBN 0-7377-2629-6 (hard cover : alk. paper)
1. Oil spills—Environmental aspects—Juvenile literature. 2. Oil spills—Cleanup—Juvenile literature. I. Title. II. Series.
TD427.P4P374 2005
628.1'6833—dc22
2004022997

Printed in the United States of America
3 4 5 6 7 12 11 10 09 08

contents

chapter one

Unnatural Disasters

People all over the world depend on oil for products they use regularly. Oil is used to make gasoline for automobiles and fuel for airplanes. It is used to make asphalt for highways and roads. Oil is also converted into chemicals used to make plastics, synthetic fibers, paint, fertilizer, and many other products. About 3 billion gallons (11 billion liters) of oil are used throughout the world *every day.* That is an enormous amount—enough to fill 3,000 school gymnasiums from floor to ceiling.

Because there is such great demand for oil, it is often transported between distant locations. For instance, much of the oil used by the United States and other countries comes from the Middle East. A large percentage of the world's oil is carried across the ocean, generally on huge tankers and barges. In most cases, these ships complete their journeys without any problems. But sometimes they have accidents. They may crash into rocks or collide with other ships or they may develop cracks or break apart during violent storms. When that happens, the vessels' oil tanks can rupture, causing massive amounts of oil to spill into the ocean.

Oil and Water

When people speak of oil spills, they are usually referring to these types of serious accidents. But oil ends up in oceans and rivers many other ways as well. For instance, some ships dump used oil (known as sludge) overboard. According to the

As the oil tanker Prestige *sinks off the Spanish coast in November 2002, millions of gallons of oil spill into the ocean.*

National Academy of Sciences (NAS), millions of gallons of **oil sludge** are deliberately dumped into the ocean each year. Oil on roads and highways can wash into streams and creeks during rainstorms. The same is true of used oil that is poured on the ground or into storm sewers after vehicle oil changes. Also, recreational watercraft such as powerboats and jet skis leak motor oil into lakes and rivers. These bodies of water channel into larger rivers that eventually flow into the ocean, and the oil is carried along with them.

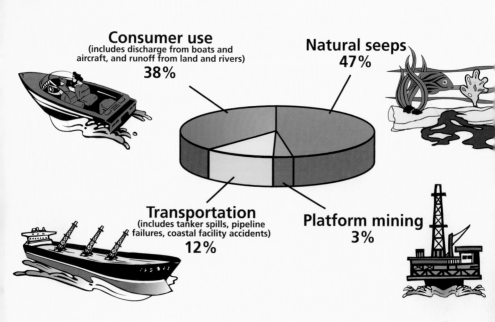

Sources of Oil in the World's Oceans

Consumer use
(includes discharge from boats and aircraft, and runoff from land and rivers)
38%

Natural seeps
47%

Transportation
(includes tanker spills, pipeline failures, coastal facility accidents)
12%

Platform mining
3%

Nature itself is responsible for leaking oil into the ocean. Crude oil is buried deep within the earth, with much of it lying beneath the sea. This oil constantly seeps into the water through cracks in the ocean floor. Scientists say that millions of gallons of oil naturally seep into the ocean every year. One well-known area where this happens is Coal Oil Point, located off the coast of Santa Barbara, California. Every day, as much as 3,000 gallons (11,000 liters) of crude oil oozes into the water from the seafloor.

Together, these sources are responsible for as much as 90 percent of the oil that ends up in the world's oceans. But since major oil spills can cause damage so quickly, they often get worldwide attention. One such incident occurred in November 2002 off the coast of Spain. An accident caused the tanker *Prestige* to spill around 20 million gallons (76 million liters) of oil into the Atlantic Ocean. Much of the oil washed ashore, and it covered hundreds of miles of beaches along the coasts of France and Spain.

Aging Ships

The *Prestige* and other supertankers are enormous. They are as long as two or three football fields. They are so big that deckhands use bicycles to get around on them. The ships are built for transporting oil. However, some are not sturdy enough to withstand strong winds and pounding ocean waves. Many of the tankers are old and have developed rust and cracks. That can make them unsafe for long ocean voyages, especially those that involve carrying millions of gallons of oil. Joseph Gross, an officer with the American Merchant Marine, has worked on many oil tankers. He says the enormous ships have problems that most people would not even consider: "Looking along the main deck," he explains, "you can actually see the vessel twist and flex. This will eventually lead to cracks in the ship's structure."[1]

The *Prestige* tanker was 26 years old, and some experts say it was not seaworthy enough to transport oil. In 2001, an inspection had discovered cracks in the ship's **hull**. The inspectors ordered repairs and the cracks were welded. One year later, when the tanker was caught in powerful storms, its hull cracked apart. Water started flooding the ship and oil began pouring out. Over the following days, the *Prestige* continued to be battered by gale-force winds and high waves. Less than a week after the accident, the tanker snapped in two. It sank to the bottom of the ocean, and most of its oil spilled into the water.

Other Contributors

People often connect massive spills like this with oil tankers. Yet even though tanker accidents have caused most of the largest oil spills, there are other causes as well. For instance, when hurricanes or earthquakes damage huge storage tanks, oil can spill out of them. Another source of oil spills is damaged pipelines. Throughout the world, there are thousands of miles of pipelines through which oil travels. Sometimes these pipes rupture and burst, which can send oil gushing into the ocean. In 2000, a pipeline broke in Rio de Janeiro, Brazil, and more than 300,000 gallons (1.1 million liters) of oil spilled into Guanabara Bay.

Another type of accident, known as a **blowout**, can happen at oil wells. Blowouts are the result of pressure that builds up inside a well. This happened in 1979, at an oil well called the *Ixtoc 1* in Mexico. For months, workers fought a raging fire and tried

Sinking tankers are not the only cause of massive spills. In 1979 a blowout at Mexico's Ixtoc 1 *oil well resulted in one of the world's worst oil disasters.*

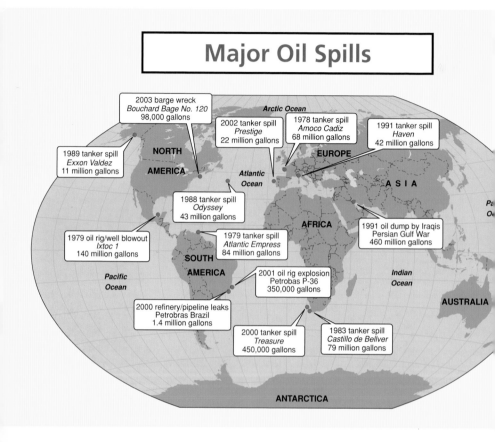

Major Oil Spills

2003 barge wreck
Bouchard Bage No. 120
98,000 gallons

Arctic Ocean

2002 tanker spill
Prestige
22 million gallons

1978 tanker spill
Amoco Cadiz
68 million gallons

1991 tanker spill
Haven
42 million gallons

1989 tanker spill
Exxon Valdez
11 million gallons

NORTH
AMERICA

EUROPE

Atlantic
Ocean

ASIA

1988 tanker spill
Odyssey
43 million gallons

AFRICA

1991 oil dump by Iraqis
Persian Gulf War
460 million gallons

1979 oil rig/well blowout
Ixtoc 1
140 million gallons

1979 tanker spill
Atlantic Empress
84 million gallons

SOUTH
AMERICA

Pacific
Ocean

Indian
Ocean

2001 oil rig explosion
Petrobas P-36
350,000 gallons

AUSTRALIA

2000 refinery/pipeline leaks
Petrobras Brazil
1.4 million gallons

2000 tanker spill
Treasure
450,000 gallons

1983 tanker spill
Castillo de Bellver
79 million gallons

ANTARCTICA

to stop the leaking oil. By the time the situation was under control, 140 million gallons (530 million liters) of oil had spilled into the Gulf of Mexico. It was one of the largest spills in history.

A Rare Occurrence

As destructive as these major spills can be, they do not happen very often. Plus, the frequency of such spills has declined over the years. Analysts say that the number of major oil spills has dropped by about two-thirds since the 1970s. This is likely due to tighter safety standards and advances in technology.

Of all the oil that ends up in the ocean each year, major spills are responsible for just a small amount. The NAS says that only about 3 percent of the oil that enters the ocean is the result of oil exploration and extraction. But according to an NAS report, there are still many risks with drilling and transporting oil. James Coleman, one of the report's authors, says that even though oil spills have decreased in the past two decades, "the potential is still there for a large spill, especially in regions with lax safety controls."[2]

As the world continues to demand more oil, greater amounts will need to be piped and transported. This is a concern because it means there is a greater risk for major oil spills. However, even as the demand has grown, such spills have grown less frequent. That is a trend scientists and environmentalists hope will continue in the future.

chapter two

The Aftermath of an Oil Spill

Whenever oil spills into a body of water, it spreads quickly, forming what is known as an oil slick. As the spreading continues, the layer grows thinner and thinner. Nature starts to react immediately. Some of the oil evaporates into the atmosphere. Then, through a process known as **weathering**, chemical and physical changes start to take place. Wind, waves, and sunlight begin breaking the oil down. Then the oil starts to **biodegrade**, which means **microorganisms** such as bacteria and fungi feed on it. Over time, this weathering causes much of the oil to disappear naturally. That is why spills in the open ocean pose less danger than those occurring close to coastlines.

Oil slicks are very unpredictable, however. They may not stay in one place long enough for nature to get rid of them. Depending on the strength of the wind, as well as tides and currents, the slicks often drift. For instance, the *Ixtoc 1* oil well blowout happened in Mexico, but the oil slick began traveling northward. Within two months it had blackened Texas coastlines hundreds of miles away.

Because oil slicks can travel so rapidly, it is important for them to be contained as soon as possible after a spill occurs. Workers must try to keep

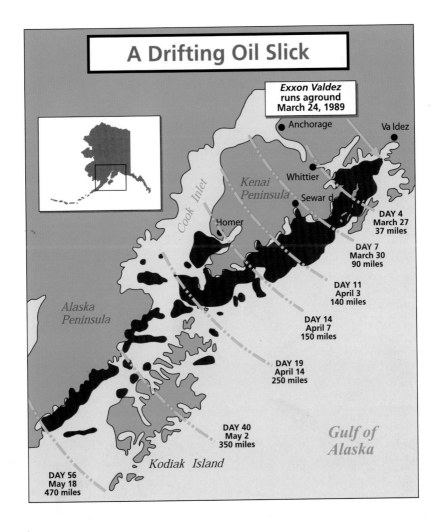

A Drifting Oil Slick

Exxon Valdez runs aground March 24, 1989

Anchorage

Valdez

Whittier

Kenai Peninsula

Seward

Cook Inlet

Homer

DAY 4
March 27
37 miles

DAY 7
March 30
90 miles

DAY 11
April 3
140 miles

Alaska Peninsula

DAY 14
April 7
150 miles

DAY 19
April 14
250 miles

DAY 40
May 2
350 miles

Gulf of Alaska

Kodiak Island

DAY 56
May 18
470 miles

the oil from reaching the coastline, where it can do the most damage. One way they do this is to surround the oil slick with floating barriers called **booms**. These devices act as fences that keep the oil confined. Once booms are in place, crews use skimmers to remove oil from the water. Skimmers may be operated from vessels or from the shoreline. One type of skimmer acts like a giant vacuum cleaner. It sucks the oil from the water and stores it in a tank, similar to how a vacuum sucks dirt into a

Alaskan fishermen place a floating boom in the water to keep an oil slick from spreading. Booms act like fences that surround and contain spilled oil.

bag. Other types float in the water and perform more like huge mops that are able to attract and hold the oil. Later, it can be squeezed into containers for recycling or disposal.

When seas are extremely rough, chemicals known as **dispersants** may be used. Dispersants are sprayed onto oil slicks by helicopters, ships, or low-flying airplanes. The chemicals inside them cause oil to break into tiny droplets, similar to the way detergent dissolves grease. Once the oil has been broken up, biodegradation causes natural bacteria to eat it.

Frothy Mousse and Tarballs

Along with keeping oil from spreading, there is another reason oil spills must be cleaned up quickly. When oil mixes with seawater, it begins to thicken, or emulsify. In the same way an eggbeater whips egg whites, waves whip the oil into a thick, frothy sludge. The sludge resembles a pudding-like substance, so it is often called "chocolate mousse." It is much more difficult to clean up than liquid oil because it cannot be vacuumed or mopped from the water. Instead, it just drifts along the surface like a fluffy, oily blanket.

Wind and waves cause the mousse to harden. Eventually, tarlike clumps form, which can drift along with currents and tides. These "tarballs" are hard and crusty on the outside and gooey on the inside. They can wash onto beaches hundreds of miles away. Although tarballs are not as damaging

When oil mixes with seawater, it thickens into a brown sludge that washes onto beaches hundreds of miles from the site of the spill.

to the environment as liquid oil, the ugly black lumps can litter beaches for many years after an accident occurs.

Deadly Coastal Spills

Despite all the efforts to contain oil before it reaches a shoreline, there are times when it is not possible. For instance, storms and rough seas can keep crews from reaching a vessel. That was the case in 1978, when a massive oil spill occurred off the coast of France. The *Amoco Cadiz* tanker developed problems with its steering. A tugboat tried towing the crippled vessel out to sea, but the water was too rough and the cable snapped. The *Cadiz* drifted toward the shoreline, and then high waves tossed it into rocks. It cracked into several pieces and spilled more than 68 million gallons (257 million liters) of oil. Although crews worked hard to keep the oil from reaching the shore, strong winds and waves hindered their efforts. As a result, a massive oil slick blackened more than 200 miles (322km) of coastline. When that happens, it is known as **black tide**.

Surf blackened with oil from the Prestige *spill crashes against rocks along the polluted Spanish coastline.*

By far, this type of coastal oil spill causes the worst environmental damage. Before the *Amoco Cadiz* accident, tourists had flocked to the French coastline to enjoy the pristine beaches. Afterward, the beaches were covered with sticky black oil. On some of the beaches, the oil was as much as 20 inches (50cm) deep. Also, fishing had been a flourishing industry in the coastal towns, but it was destroyed for several years after the spill. Biological sciences professor and author Joanna Burger describes the aftermath:

> The fishing boats remained in port and along the shores, coated in oil. Edible crabs and shore crabs, species of commercial value, lay dead and dying in the oil along the beach. Oil cov-

The 1978 Amoco Cadiz *tanker spill blackened more than 200 miles of French coastline. The disaster crippled the area's tourism and fishing industries.*

ered some two thousand acres of oyster beds in rich fishing grounds that provided a third of France's seafood. A multimillion-dollar seaweed industry was threatened. . . . The effects on other creatures were massive as well.[3]

Eleven years later, another coastal spill proved to be even more devastating. It happened in a pristine area of Alaska known as Prince William Sound. The *Exxon Valdez* supertanker ran aground on an island called Bligh Reef. Rocks tore huge holes in the ship's hull and ruptured eight of the vessel's eleven oil tanks. In less than four hours, 11 million gallons (42 million liters) of oil had spilled into the sea. The accident occurred in a remote, partially enclosed area, and the oil coated 1,300 miles (2,100km) of Alaska shoreline. Jeep Rice, a scientist from Alaska, explains why it was so destructive: "When you have an enormous oil spill in a semicontained environment like that, the oil just sloshes around and contaminates everything it touches. There is bound to be widespread devastation."[4]

Time, Nature, and Oil

To those who witness the damage of a major oil spill, it may seem as though the environment is ruined forever. But because oil itself is a natural substance, nature can heal even the most badly damaged coastlines. For instance, in the years since the *Exxon Valdez* spill, winter storms and the crashing

The Exxon Valdez *founders in Alaska's Prince William Sound as oil pours from its tanks. Floating booms contain some of the oil near the ship's stern.*

surf have cleansed Prince William Sound. Beaches that were once soaked with oil now look much like they did before the accident. There are signs, however, that more time is needed before the Alaskan shore is fully healed. In an article in *Science World* magazine, journalist Mark Bregman describes some lingering effects: "A decade after the spill, Prince William Sound's waters sparkle, and its green shores hardly betray signs of damage. But walk along a lonely march, and your footsteps might draw an oily sheen; lift a rock on a beach and you'll spot globs of oil, and oil hardened into asphalt on sandy beaches."[5]

Throughout history, nature has shown the ability to repair most anything. It has already made a great deal of progress in healing Prince William Sound. As the years go by and nature continues to perform its miracles, perhaps all traces of damage will be erased. Environmentalists hope for this not only in Alaska, but anywhere a major oil spill has caused great harm.

chapter three

Saving Wildlife

The greatest threat from an oil spill is to wildlife. Sea animals such as otters or seals can die from oil-coated fur. The oil weighs them down so they sink and drown when they try to swim. The oil also destroys their fur's insulating properties, and the animals can freeze to death. Also, animals such as sea otters and polar bears use their mouths for grooming. As they lick their fur to clean themselves, they cannot help swallowing the oil. Toxins in the oil can cause a slow, painful death by poisoning.

Seabirds are especially vulnerable to harm from oil spills. The way their feathers are arranged helps insulate the birds, which keeps them warm. The feathers also make them buoyant so they can float on the water. When oil covers the birds, their feathers mat together into clumps. This allows water to

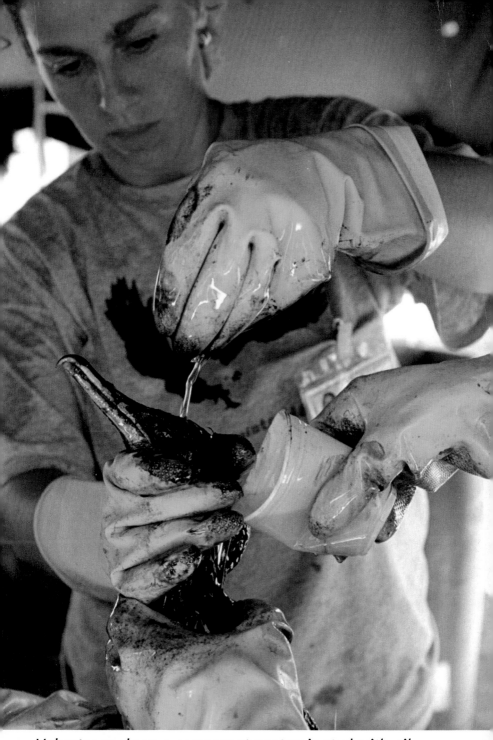

Volunteers clean a cormorant contaminated with oil. To save the bird's life, every bit of oil must be scrubbed from its feathers.

seep between the feathers and skin, causing the birds to freeze to death. Also, oil destroys the feathers' buoyancy. In an effort to keep from sinking, the birds must swim harder than usual, and they often die of exhaustion. Another risk to seabirds is poisoning. This may happen when they eat fish that have been tainted by oil. Poisoning can also occur when the birds **preen** themselves. This is the process birds use to separate and arrange their feathers with their beaks. As they preen, they can swallow oil and become poisoned.

Thousands of birds, including this cormorant, and other animals died in the wake of the Prestige *disaster.*

After the *Prestige* tanker accident, thousands of seabirds and other creatures were killed. Journalist Stephen Sackur visited one of the coastal towns affected by the spill. He said the entire coastline was coated in thick, heavy oil, and he describes what he witnessed: "The stench from the oil is truly sickening. And then, when you actually pick up the seaweed that is soaked in the stuff, you realise . . . how thick and how heavy it is and any marine life, any seabirds caught up in this simply will not stand a chance. Look closely at the waves and the rocks, and you see the black sheen everywhere."[6]

Preventive Measures

After a coastal spill, the first thing workers do is try to prevent healthy, clean creatures from coming in contact with the oil. They often use scare tactics to frighten seabirds and animals away from the area. This method was used after an April 2003 oil spill in Buzzards Bay, in Massachusetts. Crews used air horns and other loud noise-making devices to scare birds away from oil-covered beaches. At night they used brightly flashing strobe lights. Other scare tactics used to keep creatures away from spilled oil include dummies that float in the water, smoke machines, and shiny helium-filled balloons. Helicopters and boats may also be used to herd creatures away from the spill site.

Sometimes rescue workers save wildlife by capturing clean animals before the oil can touch them.

One such rescue mission occurred in June 2000. A ship called *Treasure* had sunk about 20 miles (30km) from the coast of South Africa. The area was home to thousands of African penguins, and thick oil coated the coastline. Volunteers from all over the world rushed to save the endangered birds. They were taken from their nests and carefully packed into vented boxes. The boxes were placed in helicopters and boats, and transported to other South African cities. From there the penguins were released, and they began swimming back toward their native habitat. Their journey took from ten to twenty days. By the time the birds arrived, the shoreline was safe because crews had finished cleaning up the oil. Volunteers were able to save the lives of about 40,000 penguins in the rescue operation.

Caring for Oiled Birds

Unfortunately, though, thousands more African penguins had become coated in oil before workers

African penguins are returned to the sea after volunteers saved them from the 2000 Treasure *oil spill off the South African coast.*

could reach them. Veterinarians and trained volunteers captured as many oiled birds as possible. They set up stations where the birds could be cared for. The penguins were physically examined and then given food and water. When they were strong enough to be washed, they were placed in basins of warm, soapy water. Workers scrubbed the birds to clean off every trace of oil. Jay Holcomb, of the International Bird Research Center, describes the process: "Really what we're doing is shampooing the birds," he says. "We do it vigorously, but they don't like to be handled. So we do it quickly— quickly but efficiently."[7] After the penguins had been scrubbed, they were rinsed thoroughly in warm water and placed in cages with air dryers. Then they were released into a pool of warm water. For a few weeks, workers kept a close watch to make sure the penguins were healthy and able to float. During that time, the birds preened to get their feathers back in order. The day finally came when they could survive on their own. Workers released the penguins into the ocean, and they swam back to their nests.

Saving Sea Animals

This procedure is typical after any major spill endangers wildlife. For instance, after the *Exxon Valdez* spill, a sea otter rescue center was set up and staffed by trained volunteers. More than 350 otters were taken to the center. They were given food and water, as well as medicine to absorb poisons from oil they had swallowed. They were big animals— and they were terrified. To keep them from biting workers who were trying to help them, the otters were given tranquilizers. Workers washed and

Rescuing Oiled Wildlife

Search and Collection

Workers collect live and dead animals and transport them to a care facility, counting the number of wildlife killed and injured in the incident.

Intake and Stabilization

The animal has a complete physical examination and is given medication to help it digest any oil it has accidentally swallowed. It is fed and watered up to eight times per day.

Cleaning

After two to five days of medical care and rest, the animal is washed in a series of tubs containing dishwashing detergent and hot water.

Rinsing and Drying

The animal is rinsed under high pressure, then placed in a covered pen with a pet drier until completely dry. Washing and rinsing takes about one hour per bird and several hours per otter.

Recovery

Workers monitor the animal for three to ten days to make sure its natural waterproofing is restored and all vital signs are back to normal.

Release

When the animal is healthy again, it is given a permanent identification tag and released into a clean habitat.

rinsed the sedated animals, and then put them into cages to dry. After the otters were healthy and clean enough to survive in the wild, about 200 were released. Those that were wounded or too young to survive on their own were given to aquariums.

Helpless Victims

When oil blackens a coastline, workers use many methods to clean up the mess. Because of the great risk to wildlife, saving as many creatures as possible is a high priority. Some birds and animals can be scared away before they come in contact with the oil. Others may be captured and released far from the site of the spill. But for the creatures whose fur or feathers become caked with the gooey black substance, the risk is very great. With enough time and effort, rescue workers may be able to save some of them. Many more, however, have no chance of survival—and that is, by far, the most tragic result of a coastal oil spill.

chapter Four

What Can Be Done?

Of all the oil spills that have occurred throughout history, the *Exxon Valdez* spill does not even rate among the top 50 largest. Yet none has had as great an effect on safety measures, prevention, and cleanup. When the supertanker spilled its cargo of oil in 1989, no one was prepared for such a serious accident. There was no emergency plan to handle a massive spill, and equipment was either inadequate or unavailable. By the time the cleanup operation was under way, the Alaska coastline was soaked with oil, and the situation was viewed as an environmental catastrophe.

Better Ships

Yet as tragic as the *Exxon Valdez* spill was, some positive things happened as a result of it. For instance,

Congress passed the Oil Pollution Act of 1990. It requires that all new tankers operating in U.S. waters be built with double hulls. Existing single-hulled tankers can operate in U.S. waters only until the year 2010. Unlike single-hulled ships such as the *Exxon Valdez,* double-hulled tankers have two layers of steel separating their oil tanks from the ocean. In the event that a ship's outer hull is damaged, an inner hull provides extra protection against spills.

Other countries have also passed laws requiring tankers to have double-hulled structures. For instance,

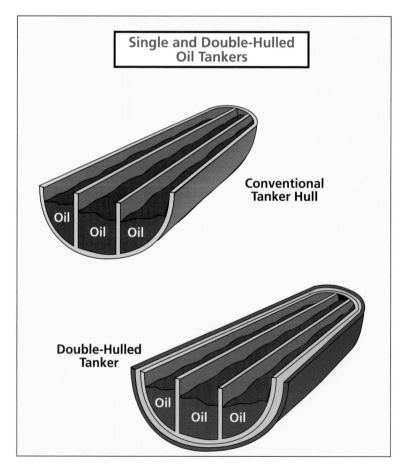

Single and Double-Hulled Oil Tankers

Conventional Tanker Hull

Oil Oil Oil

Double-Hulled Tanker

Oil Oil Oil

in the wake of the *Prestige* disaster, France and Spain now require ships traveling through their waters to be double-hulled. Many experts say the *Prestige* may not have ruptured if it had a double-hull design. In Australia, vessels owned and operated by oil companies are subject to strict requirements. New tankers have double hulls or other design improvements that reduce loss in the event of serious accidents at sea.

Prevention and Accountability

Technology also plays an important role in preventing oil spills. Many ships are now equipped with radar equipment that helps improve navigation. Marine traffic control systems are in place in shipping areas throughout the world. These guide ships in much the same way as air traffic control systems guide airplanes, and they can help decrease the risk of collisions. Also, many tankers now use high-tech equipment such as global positioning systems (GPS). In the past, tanker crews relied on two-way radios. With GPS, the Coast Guard can pinpoint a ship's exact location and monitor its progress along the way between port and sea.

Many countries punish oil companies for spilling oil in their waters. Faced with large fines and other costs, they may keep their ships in better condition. Those found guilty of negligence or misconduct must pay all cleanup costs. This can amount to millions of dollars. The oil companies may also be required to pay people who suffered losses as a result of a spill.

Disaster Drills

The best tactic is preventing oil spills from happening. Improved ship designs could dramatically reduce the number of these spills. As the demand for oil continues to grow, however, more super-tankers will travel in the world's oceans. That means there will probably always be oil spills. To prepare for such accidents, ports all over the world have put emergency plans in place. Crews attend training sessions and hold regular drills that test their ability to respond to emergencies. By holding drills, workers are better prepared to cope with a real spill.

Sometimes these drills involve pretending oil spills have occurred. For instance, during the summer of 2004, response crews in Norway pretended that oil had spilled into the North Sea. They did this in a rather unusual way—dumping a huge load of popcorn into the ocean. When the popcorn mixed with the water, it became soggy and behaved much like an oil slick. Three hundred people, thirty boats, and two aircraft participated in the drill. As the boats used booms to encircle the "spill," crews on shore practiced cleanup techniques. Kare Jorgensen, the adviser for the exercise, says popcorn is an excellent substance to use in simulating a real oil spill. If any escaped from booms, it would be harmless to the

Many countries punish companies responsible for spilling oil in their waters. The cost of cleanup efforts alone can amount to millions of dollars.

environment because it would eventually vanish. The popcorn would become another source of food for fish, birds, and other wildlife.

A similar drill was held in New Hampshire's Great Bay in 2003. But instead of popcorn, crews dumped peat moss and oranges into the water. The peat moss spread across the surface like oil, and the oranges bobbed up and down on the waves like tarballs. Barges contained the "oil slick" with 3,000 feet (914m) of booms. Skimmer boats circled in preparation to remove the simulated oil from the water. It was fortunate that this was only a drill, however, because a problem arose. One end of the boom slipped off its mooring. If oil instead of peat moss had been floating in the bay, the seabirds that populate the nearby rocks would have been in great danger. These types of exercises help response crews perfect their cleanup operations before a real emergency occurs.

Oil-Eating Chemicals

Skimmers are effective in removing oil from the water's surface. However, once the oil reaches a coastline, other cleanup methods are necessary. One type of technology, known as **bioremediation**,

Rescuers evacuate workers from an oil platform during a training exercise. Such exercises help rescue crews prepare for real-life incidents.

shows great potential for cleaning up coastal spills. Bioremediation uses microscopic organisms (such as bacteria) to speed up the rate at which natural biodegradation occurs. It is especially useful for getting rid of oil that is buried or trapped in sediments. At the time of the *Exxon Valdez* spill, bioremediation was a brand new science. Scientists applied fertilizer to certain areas of the Alaskan shoreline. Over time, the bacteria began to eat the oil, which reduced the amount on rocks and sand particles. Scientists are excited about bioremediation because it could help nature do its job faster. Whereas it could take up to ten years for a shoreline to be cleansed naturally, bioremediation could achieve the same result in two to five years.

Could Horses Fix Oil Spills?

New types of dispersants are also being developed. For instance, researchers in Scotland are testing a type of dispersant that is highly unusual. It is made from horse sweat. When a horse runs in a race, a soapy lather is produced on its pelt. The lather contains a protein that helps perspiration spread over the animal and cool it down. University of Glasgow scientists say this lather is unique because it behaves much like a household detergent. When it is applied to oil, it helps break it down. After conducting experiments, the scientists have succeeded in making a liquid that acts like horse sweat. They believe it could hold the key to dispersing oil slicks without the use of chemicals.

New technologies may one day help us to minimize the tremendous environmental impact of oil spills.

Looking Ahead

Even when every possible step is taken to prevent oil spills, accidents are going to happen. Worldwide demand for oil has grown over the past decades, which means more oil must be transported through the ocean. By putting emergency response plans in place, improving ship designs, and developing better technology, perhaps massive oil spills can be avoided. In that case, environmental disasters such as those caused by the *Exxon Valdez,* the *Prestige,* and the *Amoco Cadiz* will become nothing more than ugly reminders of the past.

Notes

Chapter 1. Unnatural Disasters

1. Joseph Gross, "Tankers and Spills," *Ethical Spectacle,* January 1996. www.spectacle.org/196/gross.html.

2. Quoted in Bill Kearney, "Millions of Gallons of Petroleum from Human Activities Enter North American Waters Annually; Most Comes from Runoff, Small Watercraft," National Academy of Sciences press release, May 23, 2002. www4.nationalacademies.org/news.nsf/isbn/0309084385?OpenDocument.

Chapter 2. The Aftermath of an Oil Spill

3. Joanna Burger, *Oil Spills.* New Brunswick, NJ: Rutgers University Press, 1997, p. 39.

4. Quoted in Stefan Lovgren, *"Exxon Valdez* Spill, 15 Years Later: Damage Lingers," *National Geographic News,* March 22, 2004. http://news.nationalgeographic.com/news/2004/03/0318_040318_exxonvaldez.html.

5. Mark Bregman, "Can Alaska Heal?" *Science World,* April 10, 2000, p. 10.

Chapter 3. Saving Wildlife

6. Stephen Sackur, "Struggle to Save a Way of Life," *BBC News,* November 19, 2002. http://news.bbc. co.uk/1/hi/world/europe/2492437.stm.

7. Quoted in John Roach, "Shampooing to Stop Oil Spill Bird Deaths," *National Geographic News,* September 21, 2004. http://news.nationalgeographic. com/news/2004/09/0921_040921_oil_penguin. html.

Glossary

biodegrade: The natural breakdown of organic material by microorganisms, such as bacteria.

bioremediation: Using microscopic organisms, such as bacteria, to speed up the rate at which natural biodegradation occurs.

black tide: A name used to describe a coastline after an oil spill.

blowout: An uncontrollable flow of fluids from a well due to high pressure inside it.

booms: Floating pieces of equipment used to contain oil after it spills into a body of water.

dispersants: Chemical substances that can break oil up into tiny droplets.

hull: The main structure of a ship (the part that is in the water).

microorganisms: Living organisms, such as bacteria, that are so small they can only be seen under a microscope.

oil sludge: Used motor oil.

preen: The process birds use to clean and arrange their feathers.

weathering: The process by which oil or other substances undergo chemical and physical changes.

For Further Exploration

Books

Nichol Bryan, *Exxon Valdez: Oil Spill.* Milwaukee: World Almanac Library, 2004. An informative account of the 1989 oil spill in Alaska's Prince William Sound.

Roland Smith, *Sea Otter Rescue: The Aftermath of an Oil Spill.* New York: Puffin Books, 1999. Discusses the details of rescuing sea otters from oil spills.

Magazines

Kim Y. Masibay, "Oil Spill! A Marine Calamity Raises the Burning Question: How to Prevent a Repeat Disaster?" *Science World,* April 18, 2003. A good article for young people that explains the November 2002 oil spill caused by the oil tanker *Prestige.*

Stephen Sawicki, "Oil Spill Renews Call for Double Hulls," *Animals,* Winter–Spring 2003. A story about increased safety measures as a result of the *Prestige* tanker oil spill in November 2002.

Internet Sources

Environment Canada, "Oil, Water, and Chocolate Mousse," 1994. www.ec.gc.ca/ee-ue/pub/chocolate/

toc_e.asp. An informative online publication about oil spills, including causes, methods of prevention, and how cleanup operations work.

U.S. Environmental Protection Agency (EPA), "Understanding Oil Spills and Oil Spill Response," December 1999. www.epa.gov/oilspill/pdfbook. htm. An excellent resource that explains oil spills in great detail.

Web Sites

Global Marine Oil Pollution Information Gateway (http://oils.gpa.unep.org/index.htm). A great site that includes facts about oil spills, effects on the environment, examples of some major spills, and what is being done to prevent them. Features a special "Kids' Pages" section.

National Oceanic and Atmospheric Administration Office of Response and Restoration (http://response.restoration.noaa.gov/index. html). Another informative site about oil pollution and oil spills, which also includes a special "For Kids" section.

National Oil Spill Response Test Facility—For Students (www.ohmsett.com/For_Students.htm). Developed for young people, this site includes frequently asked questions about oil spills, activities for students, and links. There is also an excellent description of oil spill cleanup entitled "Boomer and Dr. Skimmer Answer Questions About Oil Spill Cleanup Equipment and Methods."

U.S. Environmental Protection Agency (EPA)—Oil Program (www.epa.gov/oilspill). Provides information about the EPA's program for preventing, preparing for, and responding to oil spills in and around U.S. waterways.

Index

Picture credits

About the Author

Peggy J. Parks holds a bachelor of science degree from Aquinas College in Grand Rapids, Michigan, where she graduated magna cum laude. She is a freelance author who has written more than 30 titles for Thomson Gale's KidHaven Press, Blackbirch Press, and Lucent Books imprints. Parks lives in Muskegon, Michigan, a town she says inspires her writing because of its location on the shores of Lake Michigan.